Florian Ion **PETRESCU**

New Doppler Effect

USA 2012

Scientific reviewer:

Prof. Dr. Eng. Nicolae Mihăilescu

Copyright

Title book: New Doppler Effect

Author book: Florian Ion PETRESCU

© 2011, Florian Ion PETRESCU

petrescuflorian@yahoo.com

ALL RIGHTS RESERVED. This book contains material protected under International and Federal Copyright Laws and Treaties. Any unauthorized reprint or use of this material is prohibited. No part of this book may be reproduced or transmitted in any form or by any means, electronic or mechanical, including photocopying, recording, or by any information storage and retrieval system without express written permission from the authors / publisher.

ISBN 978-1-4699-4882-9

Welcome! A Short Book Description

The Doppler effect (or Doppler shift), named after Austrian physicist Christian Doppler who proposed it in 1842 in Prague, is the change in frequency of a wave for an observer moving relative to the source of the wave.

It is commonly heard when a vehicle sounding a siren or horn approaches, passes, and recedes from an observer.

The received frequency is higher (compared to the emitted frequency) during the approach, it is identical at the instant of passing by, and it is lower during the recession.

The relative changes in frequency can be explained as follows.

When the source of the waves is moving toward the observer, each successive wave crest is emitted from a position closer to the observer than the previous wave.

Therefore each wave takes slightly less time to reach the observer than the previous wave.

Therefore the time between the arrival of successive wave crests at the observer is reduced, causing an increase in the frequency.

While they are travelling, the distance between successive wave fronts is reduced; so the waves "bunch together".

Conversely, if the source of waves is moving away from the observer, each wave is emitted from a position farther from the observer than the previous wave, so the arrival time between successive waves is increased, reducing the frequency.

The distance between successive wave fronts is increased, so the waves "spread out".

For waves that propagate in a medium, such as sound waves, the velocity of the observer and of the source is relative to the medium in which the waves are transmitted.

The total Doppler Effect may therefore result from motion of the source, motion of the observer, or motion of the medium.

Each of these effects is analyzed separately.

For waves which do not require a medium, such as light or gravity in general relativity, only the relative difference in velocity between the observer and the source needs to be considered.

The Doppler Effect [1-3] represents the frequency variation of the waves, received by an observer which is drawing (coming), respectively it's removing (going), from a wave spring (source).

If a bright spring is drawing to an observer, the frequency of waves received by the observer is bigger than the emitted frequency of source, such that the respective spectral lines are moving to violet.

On the contrary, if the light source is removing from the observer, the spectral lines are moving to red.

One proposes to study the Doppler Effect for the light waves, generally for the electromagnetic waves.

CHAPTER I - ABOUT THE DOPPLER EFFECT

Introduction

The Doppler effect (or Doppler shift), named after Austrian physicist Christian Doppler who proposed it in 1842 in Prague, is the change in frequency of a wave for an observer moving relative to the source of the wave.

It is commonly heard when a vehicle sounding a siren or horn approaches, passes, and recedes from an observer.

The received frequency is higher (compared to the emitted frequency) during the approach, it is identical at the instant of passing by, and it is lower during the recession (See the Figure 1).

The relative changes in frequency can be explained as follows.

When the source of the waves is moving toward the observer, each successive wave crest is emitted from a position closer to the observer than the previous wave.

Therefore each wave takes slightly less time to reach the observer than the previous wave.

Therefore the time between the arrival of successive wave crests at the observer is reduced, causing an increase in the frequency.

While they are travelling, the distance between successive wave fronts is reduced; so the waves "bunch together".

Conversely, if the source of waves is moving away from the observer, each wave is emitted from a position farther from the observer than the previous wave, so the arrival time between successive waves is increased, reducing the frequency.

The distance between successive wave fronts is increased, so the waves "spread out".

For waves that propagate in a medium, such as sound waves, the velocity of the observer and of the source is relative to the medium in which the waves are transmitted.

The total Doppler Effect may therefore result from motion of the source, motion of the observer, or motion of the medium.

Each of these effects is analyzed separately.

For waves which do not require a medium, such as light or gravity in general relativity, only the relative difference in velocity between the observer and the source needs to be considered.

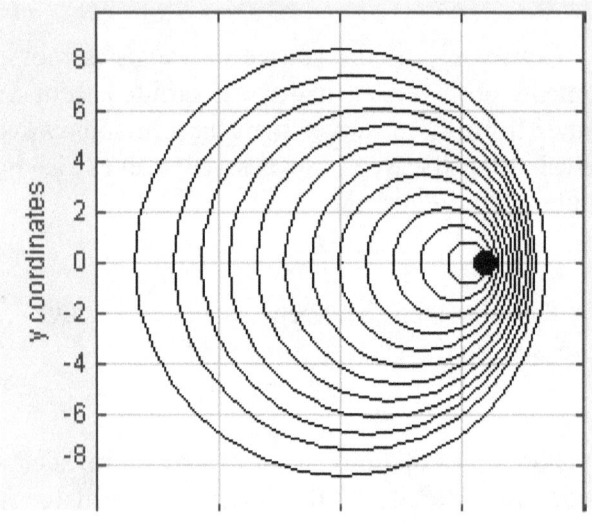

Fig. 1 *The Doppler Effect model*

Development

Doppler first proposed the effect in 1842 in his treatise "Über das farbige Licht der Doppelsterne und einiger anderer Gestirne des Himmels" (On the coloured light of the binary stars and some other stars of the heavens).

The hypothesis was tested for sound waves by Buys Ballot in 1845.

He confirmed that the sound's pitch was higher than the emitted frequency when the sound source approached him, and lower than the emitted frequency when the sound source receded from him.

Hippolyte Fizeau discovered independently the same phenomenon on electromagnetic waves in 1848 (in France, the effect is sometimes called "l'Effet Doppler-Fizeau" but that name was not adopted by the rest of the world as Fizeau's discovery was three years after Doppler's).

In Britain, John Scott Russell made an experimental study of the Doppler Effect (1848).

Craig Bohren pointed out in 1991 that some physics textbooks erroneously state that the observed frequency increases as the object approaches an observer and then decreases only as the object passes the observer.

In most cases, the observed frequency of an approaching object declines monotonically from a value above the emitted frequency, through a value equal to the emitted frequency when the object is closest to the observer, and to values increasingly below the emitted frequency as the object recedes from the observer.

Bohren proposed that this common misconception might occur because the intensity of the sound increases as an object approaches an observer and decreases once it passes and recedes from the observer and that this change in intensity is misperceived as a change in frequency.

Higher sound pressure levels make for a small decrease in perceived pitch in low frequency sounds, and for a small increase in perceived pitch for high frequency sounds.

Application

Sirens

The siren on a passing emergency vehicle will start out higher than its stationary pitch, slide down as it passes, and continue lower than its stationary pitch as it recedes from the observer.

Astronomer John Dobson explained the effect thus:

"The reason the siren slides is because it doesn't hit you."

In other words, if the siren approached the observer directly, the pitch would remain constant (as $v_{s,r}$ is only the radial component) until the vehicle hit him, and then immediately jump to a new lower pitch.

Because the vehicle passes by the observer, the radial velocity does not remain constant, but instead varies as a function of the angle between his line of sight and the siren's velocity:

$$v_r = v_s \cdot \cos\theta$$

where v_s is the velocity of the object (source of waves) with respect to the medium, and θ is the angle between the object's forward velocity and the line of sight from the object to the observer.

Astronomy

Redshift of spectral lines in the optical spectrum of a supercluster of distant galaxies (right), as compared to that of the Sun (left)

The Doppler Effect for electromagnetic waves such as light is of great use in astronomy and results in either a so-called red shift or blue shift. It has been used to measure the speed at which stars and galaxies are approaching or receding from us, that is, the radial velocity.

This is used to detect if an apparently single star is, in reality, a close binary and even to measure the rotational speed of stars and galaxies.

The use of the Doppler Effect for light in astronomy depends on our knowledge that the spectra of stars are not continuous.

They exhibit absorption lines at well defined frequencies that are correlated with the energies required to excite electrons in various elements from one level to another.

The Doppler Effect is recognizable in the fact that the absorption lines are not always at the frequencies that are obtained from the spectrum of a stationary light source.

Since blue light has a higher frequency than red light, the spectral lines of an approaching astronomical light source exhibit a blue shift and those of a receding astronomical light source exhibit a redshift.

Among the nearby stars, the largest radial velocities with respect to the Sun are +308 km/s (BD-15°4041, also known as LHS 52, 81.7 light-years away) and -260 km/s (Woolley 9722, also known as Wolf 1106 and LHS 64, 78.2 light-years away).

Positive radial velocity means the star is receding from the Sun, negative that it is approaching.

Temperature measurement

Another use of the Doppler Effect, which is found mostly in plasma physics and astronomy, is the estimation of the temperature of a gas (or ion temperature in a plasma) which is emitting a spectral line.

Due to the thermal motion of the emitters, the light emitted by each particle can be slightly red or blue-shifted, and the net effect is a broadening of the line.

This line shape is called a Doppler profile and the width of the line is proportional to the square root of the temperature of the emitting species, allowing a spectral line (with the width dominated by the Doppler broadening) to be used to infer the temperature.

Radar

The Doppler Effect is used in some types of radar, to measure the velocity of detected objects.

A radar beam is fired at a moving target — e.g. a motor car, as police use radar to detect speeding motorists — as it approaches or recedes from the radar source.

Each successive radar wave has to travel farther to reach the car, before being reflected and re-detected near the source.

As each wave has to move farther, the gap between each wave increases, increasing the wavelength.

In some situations, the radar beam is fired at the moving car as it approaches, in which case each successive wave travels a lesser distance, decreasing the wavelength.

In either situation, calculations from the Doppler Effect accurately determine the car's velocity.

Moreover, the proximity fuze, developed during World War II, relies upon Doppler radar to explode at the correct time, height, distance, etc.

Medical imaging and blood flow measurement

Color flow ultrasonography (Doppler) of a carotid artery - scanner and screen

An echocardiogram can, within certain limits, produce accurate assessment of the direction of blood flow and the velocity of blood and cardiac tissue at any arbitrary point using the Doppler Effect.

One of the limitations is that the ultrasound beam should be as parallel to the blood flow as possible.

Velocity measurements allow assessment of cardiac valve areas and function, any abnormal communications between the left and right side of the heart, any leaking of blood through the valves (valvular regurgitation), and calculation of the cardiac output.

Contrast-enhanced ultrasound using gas-filled microbubble contrast media can be used to improve velocity or other flow-related medical measurements.

Although "Doppler" has become synonymous with "velocity measurement" in medical imaging, in many cases it is not the frequency shift (Doppler shift) of the received signal that is measured, but the phase shift (when the received signal arrives).

Velocity measurements of blood flow are also used in other fields of medical ultrasonography, such as obstetric ultrasonography and neurology.

Velocity measurement of blood flow in arteries and veins based on Doppler Effect is an effective tool for diagnosis of vascular problems like stenosis.

Flow measurement

Instruments such as the laser Doppler velocimeter (LDV), and acoustic Doppler velocimeter (ADV) have been developed to measure velocities in a fluid flows.

The LDV emits a light beam and the ADV emits an ultrasonic acoustic burst, and measure the Doppler shift in wavelengths of reflections from particles moving with the flow.

The actual flow is computed as a function of the water velocity and phase.

This technique allows non-intrusive flow measurements, at high precision and high frequency.

Velocity profile measurement

Developed originally for velocity measurements in medical applications (blood flow), Ultrasonic Doppler Velocimetry (UDV) can measure in real time complete velocity profile in almost any liquids containing particles in suspension such as dust, gas bubbles, emulsions.

Flows can be pulsating, oscillating, laminar or turbulent, stationary or transient.

This technique is fully non-invasive.

Satellite communication

Fast moving satellites can have a Doppler shift of dozens of kilohertz relative to a ground station.

The speed, thus magnitude of Doppler Effect, changes due to earth curvature.

Dynamic Doppler compensation, where the frequency of a signal is changed multiple times during transmission, is used so the satellite receives a constant frequency signal.

Underwater acoustics

In military applications the Doppler shift of a target is used to ascertain the speed of a submarine using both passive and active sonar systems.

As a submarine passes by a passive sonobuoy, the stable frequencies undergo a Doppler shift, and the speed and range from the sonobuoy can be calculated.

If the sonar system is mounted on a moving ship or another submarine, then the relative velocity can be calculated.

A sonobuoy (a portmanteau of sonar and buoy; see the Figure 2) is a relatively small (typically 5 inches / 13 centimeters, in diameter and 3 ft/91 cm long) expendable sonar system that is dropped/ejected from aircraft or ships conducting anti-submarine warfare or underwater acoustic research.

The buoys are ejected from aircraft in canisters and deploy upon water impact.

An inflatable surface float with a radio transmitter remains on the surface for communication with the aircraft, while one or more hydrophone sensors and stabilizing equipment descend below the surface to a selected depth that is variable, depending on environmental conditions and the search pattern.

The buoy relays acoustic information from its hydrophone(s) via UHF/VHF radio to operators onboard the aircraft.

Fig. 2 *Sonobuoy being loaded onto an USN P-3C Orion aircraft*

Vibration measurement

A laser Doppler vibrometer (LDV) is a non-contact method for measuring vibration.

The laser beam from the LDV is directed at the surface of interest, and the vibration amplitude and frequency are extracted from the Doppler shift of the laser beam frequency due to the motion of the surface.

A laser Doppler vibrometer (LDV) is a scientific instrument that is used to make non-contact vibration measurements of a surface.

The laser beam from the LDV is directed at the surface of interest, and the vibration amplitude and frequency are extracted from the Doppler shift of the laser beam frequency due to the motion of the surface.

The output of an LDV is generally a continuous analog voltage that is directly proportional to the target velocity component along the direction of the laser beam.

Some advantages of an LDV over similar measurement devices such as an accelerometer are that the LDV can be directed at targets that are difficult to access, or that may be too small or too hot to attach a physical transducer.

Also, the LDV makes the vibration measurement without mass-loading the target, which is especially important for MEMS devices.

The relativistic Doppler Effect

The relativistic Doppler Effect (Figure 3) is the change in frequency (and wavelength) of light, caused by the relative motion of the source and the observer (as in the classical Doppler Effect), when taking into account effects described by the special theory of relativity.

The relativistic Doppler Effect is different from the non-relativistic Doppler Effect as the equations include the time dilation effect of special relativity and do not involve the medium of propagation as a reference point.

They describe the total difference in observed frequencies and possess the required Lorentz symmetry.

Fig. 3 *A source of light waves moving to the right with velocity 0.7c. The frequency is higher on the right, and lower on the left.*

The photoacoustic Doppler Effect

The photoacoustic Doppler Effect, as its name implies, is one specific kind of Doppler Effect, which occurs when an intensity modulated light wave induces a photoacoustic wave on moving particles with a specific frequency.

The observed frequency shift is a good indicator of the velocity of the illuminated moving particles. A potential biomedical application is measuring blood flow.

Specifically, when an intensity modulated light wave is exerted on a localized medium, the resulting heat can induce an alternating and localized pressure change.

This periodic pressure change generates an acoustic wave with a specific frequency.

Among various factors that determine this frequency, the velocity of the heated area and thus the moving particles in this area can induce a frequency shift proportional to the relative motion.

Thus, from the perspective of an observer, the observed frequency shift can be used to derive the velocity of illuminated moving particles.

CHAPTER II – SOME FEW SPECIFICATIONS ABOUT THE DOPPLER EFFECT TO THE ELECTROMAGNETIC WAVES

Introduction

The Doppler Effect [1-3] represents the frequency variation of the waves, received by an observer which is drawing (coming), respectively it's removing (going), from a wave spring (source).

If a bright spring is drawing to an observer, the frequency of waves received by the observer is bigger than the emitted frequency of source, such that the respective spectral lines are moving to violet.

On the contrary, if the light source is removing from the observer, the spectral lines are moving to red.

One proposes to study the Doppler Effect for the light waves, generally for the electromagnetic waves.

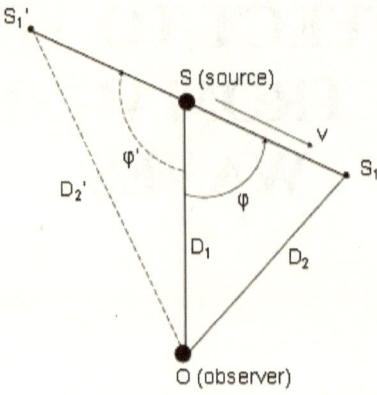

Fig. 4. *The waves received by an observer O from a waves source S, which is moving in relation with the observer, by the direction SS_1*

The new relations

We wish to calculate the period (T [s]) of the waves received by an observer O (the Fig. 4) from a waves

24

source S, which is moving in relation with the observer, on the direction SS_1 with the relative speed v [m/s] [1, 2].

T_0 [s] is the period of waves emitted by the source S.

At the moment t_0 [s], determinate by the observer O, from the source S bend a bright wave; this wave traverse the distance D_1=SO [m] and arrive in O at the moment t_1 [s].

$$t_1 = t_0 + \frac{D_1}{c} \qquad (1)$$

Where c is the light speed in vacuum: $c \cong 3 \cdot 10^8$ [m/s].

After a T_0 period, from the source S (arrived now in S_1), from the source S_1 starts a second wave. The distance SS_1 [m] is:

$$SS_1 = v.T_0 \qquad (2)$$

The observer O, receive the second waves at the moment t_2 [s]:

$$t_2 = t_0 + T_0 + \frac{D_2}{c} \qquad (3)$$

The period T is equal with the difference between the two moments.

$$T = t_2 - t_1 = T_0 + \frac{D_2 - D_1}{c} \qquad (4)$$

The angle φ [rad] between the two vectors, SS_1 and SO is known and the distance $D_1 = SO$ is known as well. With the COS theorem in the certain triangle SOS_1 one obtains the distance D_2 [m]:

$$D_2 = \sqrt{D_1^2 + SS_1^2 - 2.D_1.SS_1.\cos\varphi} \qquad (5)$$

With SS_1 from (2) the relation (5), become the expression (6).

$$D_2 = \sqrt{D_1^2 + v^2.T_0^2 - 2.D_1.v.T_0.\cos\varphi} \qquad (6)$$

With the expression (6) in relation (4) one obtains the form (7).

$$T = T_0 + \frac{\sqrt{D_1^2 + v^2 T_0^2 - 2D_1 v T_0 \cos\varphi} - D_1}{c} \qquad (7)$$

The relation (7) can be put in the form (8).

$$T = T_0 (1+\beta \frac{v.T_0 - 2.D_1.\cos\varphi}{\sqrt{D_1^2 + v^2 T_0^2 - 2D_1 v T_0 \cos\varphi} + D_1}) \qquad (8)$$

Where β is the ratio between the two speeds, v and c:

$$\beta = \frac{v}{c} \qquad (9)$$

Presents the classical relation (10)

The classical relation (10) is very simply, but it's an approximate relation [2-3].

The expression (8) is more difficult but it's a very exact relation. It can be put in the forms (18), (19) and finally (20).

$$\frac{T}{T_0} = 1 \pm \beta . \cos\varphi \qquad (10)$$

Some aspects

a) When the source S is removing from the observer, the angle φ (see the figure 1) take the values (φ') comprised between 90^0 and 180^0, cosφ become negative, the numerator of expression (8) become positive and the period of observer O (T) it'll be always bigger than T_0 (the period of source): $T>T_0$ and $y<y_0$ (the spectral lines are red).

When the source S is drawing to the observer, the angle $\varphi \in [0^0, 90^0)$ and cosφ>0. In this case one analyzes (11) the numerator of expression (8) and one can have two cases (b and c) [1]:

$$N = v.T_0 - 2.D_1.\cos\varphi \qquad (11)$$

b) If N<0, then $vT_0 < 2D_1 \cdot cos\varphi$ or

$$\cos\varphi > \frac{v \cdot T_0}{2 \cdot D_1} \qquad (12)$$

and $T<T_0$, or $y>y_0$ (the spectral lines are violet) [1].

c) If N>0, then

$$\cos\varphi < \frac{v \cdot T_0}{2 \cdot D_1} \qquad (13)$$

and $T>T_0$, or $y<y_0$ (the spectral lines are red).

This case it wasn't known by the classical expression (10) [1].

d) The most interesting case is then when the angle $\varphi=90^0$, and $\cos\varphi=0$, when the source is moving perpendicular at the axle SO (see the figure 5). In this case the relation (8), become the expression (14).

$$T = T_0(1+\frac{\beta.v.T_0}{\sqrt{D_1^2+v^2.T_0^2}+D_1}) \qquad (14)$$

$T>T_0$ and $y<y_0$ (the spectral lines are red) [1].

This fact can't be seen by the classical relation (10) which (for the $\varphi=90^0$), takes the form (15):

$$T = T_0 \qquad (15)$$

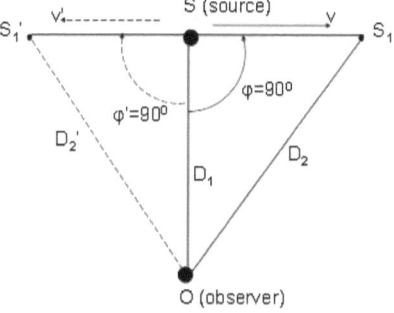

Fig. 5. *The waves received by an observer O from a waves source S when the source is moving perpendicular at the axle SO (it is a particular situation)*

The classical approximate relation (10, form 15) can't foresee the Doppler Effect for this case, but the effect virtually exist, and for this reason it was introduced the relativity effect (or the Lorentz transformation), where the period T_0 takes the form T_0/α (see [1]), and the relation (15) takes the form (16) [2, 3]:

$$T = \frac{T_0}{\alpha} \qquad (16)$$

$$\alpha = \sqrt{1-\beta^2} \qquad (17)$$

If v<c, the expression $\sqrt{D_1^2 + v^2 \cdot T_0^2 - 2 \cdot D_1 \cdot v \cdot T_0 \cdot \cos\varphi}$ =>D and the relation (8) can be approximated by the expression (18), (8=>18):

$$\frac{\Upsilon_0}{\Upsilon} = \frac{T}{T_0} = 1 - \beta \cdot \cos\varphi + \beta \cdot \frac{v \cdot T_0}{2 \cdot D_1} \qquad (18)$$

The distance D (D_1) can take different values for the same frequency Υ_0 (One can't determine D from 8 or 18; D is indeterminate. Practically, the frequency Υ is a real function of Υ_0 and; Υ is a function of Υ_0, T_0, or $\lambda_0 = c \cdot T_0$;

The distance D can't take any value. It must be a multiple of λ_0).

The relation (18) takes mandatory the forms (19) for a quantum distance ($D_1 = n \cdot c \cdot T_0$) and (20) when n takes mandatory the basic value (n=1) to keep the own original wave (one utilize just the basic frequency for n=1, see the final relation 20; for other frequencies then we can already speak about other waves):

$$\frac{\gamma_0}{\gamma} = \frac{T}{T_0} = 1 - \beta \cdot \cos\varphi + \frac{1}{2} \cdot \beta^2 \cdot \frac{1}{n} \quad (19)$$

$$\frac{\gamma_0}{\gamma} = \frac{T}{T_0} = 1 - \beta \cdot \cos\varphi + \frac{1}{2} \cdot \beta^2 \quad (20)$$

First, the relation (20) can be utilized to determine the period T when one know the source period T_0 and the source velocity, v ($\beta = v/c$).

It can speak now about a quantum Doppler Effect relation (20).

Second, if one know the two frequencies (Υ, Υ_0), one can determine the source velocity v in relation of the observer (β and $v = \beta \cdot c$), with the new relation (20) or more rapidly with the classical form (10).

Conclusion

In this work one proposes to exchange the classical relation (10) (see [1], p. 114) with the new and more exactly relation (20).

$$\frac{\gamma_0}{\gamma} = \frac{T}{T_0} = 1 - \beta.\cos\varphi \qquad (10)$$

$$\frac{\gamma_0}{\gamma} = \frac{T}{T_0} = 1 - \beta \cdot \cos\varphi + \frac{1}{2} \cdot \beta^2 \qquad (20)$$

Bibliography

[1] **Bărbulescu N.**, *"Bazele fizice ale relativității Einsteiniene"*. Editura Științifică și Enciclopedică, București, 1979, p. 142-148;
[2] **David Halliday, Robert, R.**, - *Physics, Part II*, Edit. John Wiley & Sons, Inc. - New York, London, Sydney, 1966;
[3] **Petrescu-Prahova, M., Petrescu-Prahova, I.**, - *Fizica-Manual pentru anul IV liceu, secția reală*, Editura Didactică și Pedagogică, București, 1976.

CHAPTER III – SOME SPECIFICATIONS ABOUT THE MODERN DOPPLER EFFECT

Doppler Effect Found Even at Molecular Level -- 169 Years After Its Discovery

Science Daily (May 10, 2011) — Whether they know it or not, anyone who's ever gotten a speeding ticket after zooming by a radar gun has experienced the Doppler Effect -- a measurable shift in the frequency of radiation based on the motion of an object, which in this case is your car doing 45 miles an hour in a 30-mph zone.

But for the first time, scientists have experimentally shown a different version of the Doppler Effect at a much, much smaller level -- the rotation of an individual molecule. Prior to this such an effect had been theorized, but it took a complex experiment with a synchrotron to prove it's for real.

"Some of us thought of this some time ago, but it's very difficult to show experimentally," said T. Darrah Thomas, a professor emeritus of chemistry at Oregon State University and part of an international research team that just announced its findings in Physical Review Letters, a professional journal.

Most illustrations of the Doppler Effect are called "translational," meaning the change in frequency of light or sound when one object moves away from the other in a straight line, like a car passing a radar gun. The basic concept has been understood since an Austrian physicist named Christian Doppler first proposed it in 1842.

But a similar effect can be observed when something rotates as well, scientists say.

"There is plenty of evidence of the rotational Doppler Effect in large bodies, such as a spinning planet or galaxy," Thomas said. "When a planet rotates, the light coming from it shifts to higher frequency on the side spinning toward you and a lower frequency on the side spinning away from you. But this same basic force is at work even on the molecular level."

In astrophysics, this rotational Doppler Effect has been used to determine the rotational velocity of things such as planets. But in the new study, scientists from Japan, Sweden, France and the United States provided the first experimental proof that the same thing happens even with molecules.

At this tiny level, they found, the rotational Doppler Effect can be even more important than the linear motion of the molecules, the study showed.

The findings are expected to have application in a better understanding of molecular spectroscopy, in which the radiation emitted from molecules is used to study their makeup and chemical properties. It is also relevant to the study of high energy electrons, Thomas said.

"There are some studies where a better understanding of this rotational Doppler Effect will be important," Thomas said. "Mostly it's just interesting. We've known about the Doppler Effect for a very long time but until now have never been able to see the rotational Doppler Effect in molecules."

References

[1] T. D. Thomas, E. Kukk, K. Ueda, T. Ouchi, K. Sakai, T. X. Carroll, C. Nicolas, O. Travnikova, and C. Miron. **Experimental observation of rotational Doppler broadening in a molecular system**. *Physical Review Letters*, Accepted Apr 12, 2011.

[2] Oregon State University (2011, May 10). Doppler Effect found even at molecular level -- 169 years after its discovery. *Science Daily*.
http://www.sciencedaily.com/releases/2011/05/110510134112.htm

Did Scientists Break the Speed of Light?

CERN Records Sub-Atomic Particle Speeds

September 23, 2011 - Albert Einstein is responsible for many of the longest-standing laws of physics, including the famous theory of special relativity. Nearly all of modern physics and astronomy is based upon the concept that nothing can travel faster than the speed of light. It appears that Einstein may have been proven wrong thanks to a new study completed by CERN (the European Organization for Nuclear Research).

CERN claims that they have recorded sub-atomic particles traveling at a speed faster than light. The OPERA experiment consisted of firing 15,000 beams of neutrinos from Geneva, Switzerland to Gran Sasso, Italy over a period of 3 years. The sensors in Italy registered that the particles reached the target 60 nanoseconds faster than the speed of light. According to modern knowledge, this is an impossible feat. CERN scientists asked for others to confirm their research by reproducing the results, and Fermilab in Chicago is already attempting to recreate the experiment.

Despite asking for a double check, the CERN scientists seem very sure of this finding. "We have high confidence in our results, stated spokesman Antonio Ereditato.

"We have checked and rechecked for anything that could have distorted our measurements but we found nothing." If these findings are indeed accurate, they will shake the very foundation of Einstein's theory.

While some people might think that 60 nanoseconds is insignificant, that little amount may have serious implications in the possibilities of light speed travel or even time travel. However, the neutrino is still quite mysterious to scientists, so further research is necessary to prove exactly what the OPERA experiment proved.

It's pretty amazing how little we actually know about our universe. Something new is often discovered that completely changes our perceptions of science and reality. I'm really looking forward to seeing what the scientific community discovers if the light speed barrier is actually broken.

References

[1] http://www.chacha.com/topic/light-speed/gallery/995/did-scientists-break-the-speed-of-light

Meteorologists Invent Better Way to Monitor Hurricane Strength

September 1, 2007 — Meteorologists have developed a new method for analyzing hurricane strength. A series of mathematical formulas transform data from Doppler radars into a 3-D picture of storm intensity every 6 minutes. Because of the rapid updates, the technique increases meteorologists' ability to capture sudden, dangerous changes in hurricanes.

The strongest hurricane to hit the U.S. in more than a decade -- killing ten people -- causing thirteen-billion dollars in damage. Its arrival was expected. Its intensity: an absolute surprise.

We are in the middle of hurricane season again and meteorologists are rushing to test a new way to track a storm's intensity. Scientists now know, as hurricane Charlie approached Florida three years ago, Floridians were preparing for the storm with obsolete information.

Charlie landed with 25-percent more intensity than predicted. It's a scenario that could forever be avoided with a new tracking system.

'[Hurricane Charlie] rapidly intensified from category two to category four in roughly three hours,' said Wen-Chau Lee, NCAR meteorologist.

If only meteorologists knew then what they know now. Now, meteorologists at the national center for atmospheric research have a new software tool called "VORTRAC." It slices through approaching hurricanes to reveal a three-dimensional view of the storm and just how intense it will be. The result looks a lot like the layers of a sliced onion.

'When you cut an onion in half you see different rings. Basically what we do, we dissect a hurricane into different rings,' Lee said.

VORTRAC combines wind measurements from the Doppler radar closest to the eye of the storm with existing hurricane data to show a 3-D view of the wind. Lee said he looks forward to putting his tracking system to the test in the U.S. when the next hurricane heads our way.

Because of the limited range of Doppler radars, VORTRAC works only for hurricanes within about 120 miles of land. But that could help monitor the critical 10 to 15 hours before landfall. The National Hurricane Center is testing the system currently and expects it to be ready for use in about two years.

BACKGROUND: Forecasters are testing a new technique called VORTRAC -- Vortex Objective Radar Tracking and Circulation -- that provides a detailed 3D view of an approaching hurricane every six minutes and allows them to determine whether the storm is gathering strength as it nears land. Then they can quickly alert coastal communities if it suddenly strengthens.

HOW IT WORKS: Developed by researchers at the National Center for Atmospheric Research (NCAR), the technique relies on the existing network of Doppler radars along the southeast coast to closely monitor hurricanes winds. Any radar can measure winds blowing toward or away from it, but no single radar could provide a 3D picture of hurricane winds until now.

The NCAR scientists developed a series of mathematical formulas that combine data from a single radar near the center of a landfalling storm with general knowledge of Atlantic hurricane structure in order to map the approaching system's winds in three dimensions. The technique also infers the barometric pressure in the eye of the hurricane, a very reliable index of its strength. However, because of the limited range of Doppler radars, VORTRAC works only for hurricanes that are within about 120 miles of land. In the future, it might be possible to use VORTRAC to help improve long-range hurricane forecasts by using data from airborne radars to glean detailed information about a hurricane that is far out to sea.

ABOUT HURRICANES: A hurricane is a type of tropical cyclone, a low-pressure system that usually forms in the tropics and has winds that circulate counterclockwise near the earth's surface. Storms are considered hurricanes when their wind speeds surpass 74 MPH. Every hurricane arises from the combination of warm water and moist warm air. Tropical thunderstorms drift out over warm ocean waters and encounter winds coming in from near the equator. Warm, moist air from the ocean surface rises rapidly, encounters cooler air, and condensed into water vapor to form storm clouds, releasing heat in the process.

This heat causes the condensation process to continue, so that more and more warm moist air is drawn into the developing storm, creating a wind pattern that spirals around the relatively calm center, or eye, of the storm, much like water swirling down a drain. The winds keep circling and accelerating to form a classic cyclone pattern.

WHAT IS DOPPLER RADAR: Doppler radar uses a well-known effect of light called the Doppler shift. Just as a train whistle will sound higher as it approaches a platform and then become lower in pitch as it moves away, light emitted by a moving object is perceived to increase in frequency (a blue shift) if it is moving toward the observer; if the object is moving away from us, it will be shifted toward the red end of the spectrum. Doppler radar sends out radio waves that bounce off objects in the air, such as raindrops or snow crystals, and then measures how much the frequency changes in returning radio waves to better determine wind direction and speed.

References

[1] 3D Hurricane Tracking. *Science Daily*.

http://www.sciencedaily.com/videos/2007/0901-3d_hurricane_tracking.htm

3D Doppler Ultrasound Helps Identify Breast Cancer

ScienceDaily (Oct. 21, 2008) — Three-dimensional (3-D) power Doppler ultrasound helps radiologists distinguish between malignant and benign breast masses, according to a new study being published in the November issue of Radiology.

"Using 3-D scans promises greater accuracy due to more consistent sampling over the entire tumor," said lead author, Gerald L. LeCarpentier, Ph.D., assistant professor in the Department of Radiology at University of Michigan in Ann Arbor. "Our study shows that 3-D power Doppler ultrasound may be useful in the evaluation of some breast masses."

Malignant breast masses often exhibit increased blood flow compared to normal tissue or benign masses. Using 3-D power Doppler ultrasound, radiologists are able to detect vessels with higher flow speeds, which likely indicate cancer.

For the study, Dr. Le Carpentier and colleagues studied 78 women between the ages of 26 and 70 who were scheduled for biopsy of a suspicious breast mass. Each of the women underwent a 3-D Doppler ultrasound exam followed by core or excisional biopsy of the breast.

The results showed that 3-D power Doppler ultrasound was highly accurate in identifying malignant breast tumors. When combined with age-based assessment and gray scale visual analysis, 3-D Doppler showed a sensitivity of 100 percent in identifying cancerous tumors and a specificity of 86 percent in excluding benign tumors.

"Using speed-weighted 3-D power Doppler ultrasound, higher flow velocities in the malignant tumor-feeding vessels may be detected, whereas vessels with slower flow velocities in surrounding benign masses may be excluded," Dr. Le Carpentier said.

References

[1] 3-D Doppler Ultrasound Helps Identify Breast Cancer. *Science Daily*.
http://www.sciencedaily.com/releases/2008/10/081021093933.htm

[2] Le Carpentier et al. Suspicious Breast Lesions: Assessment of 3D Doppler US Indexes for Classification in a Test Population and Fourfold Cross-Validation Scheme. Radiology, 2008; 249 (2): 463 DOI: 10.1148/radiol.2492060888

Perfusion In Burn Injuries Rapidly Determined By Using Improved Laser-Doppler Technology, Hospital Test Shows

ScienceDaily (Dec. 16, 2007) — The perfusion of a burn injury can now rapidly be determined by using a new technique developed by scientists of the University of Twente. Using the perfusion image made by a laser and an ultra fast camera, doctors will be able to determine whether a burn needs surgery. The new measuring device, developed under supervision of Dr. Wiendelt Steenbergen of the Biophysical Engineering group, has been successfully tested at the hospital Martini Ziekenhuis in Groningen.

Tests in hospital show that the system is perfectly capable of measuring differences in perfusion in burn wounds; patients and medical staff are positive about the high speed of the system and the quality of the images.

A burn that shows good perfusions, has a better chance of healing by itself: no surgery is needed. In many cases, the visual inspection is not sufficient to take a decision on the necessity of surgery. This can lead to unnecessary surgery or, on the other hand, to unwanted delays when surgery is the best option.

Compared to current perfusion measurements, the new technique is much faster. Scanning techniques take minutes of time for some square centimeters of skin, during which time the patient is not allowed to move. The new technique will be capable of imaging an entire surface of ten by ten centimeter in a fraction of a second.

Doppler Effect

In order to reach this high speed, the entire surface is lit at once using a wide laser beam. A high speed camera, capable of taking 27000 shots per second, takes images of the tissue. Whenever laser light is scattered by moving rood blood cells, this is visible in the intensity of the pixels; due to the Doppler effect, a color shift will be visible. From the resulting 'movie' of the tissue, a perfusion image can be made.

Apart from this promising application in determining perfusion in burn injuries, Wiendelt Steenbergen predicts other applications, for example in evaluating the uptake of medication through the skin, or in testing allergic reactions. In evaluating diabetic micro circulation problems, the new technique could be an attractively fast alternative to current approaches as well.

Patent

Current market leader in Laser-Doppler equipment for perfusion imaging, Perimed AB from Sweden, has shown interest in the new technique. The Swedish company has signed a contract with Dutch Technology Foundation STW, for acquiring the patent on the technique. STW finances Steenbergen's work.

The research has been done within the Biophysical Engineering Group of the University of Twente, which is part of the BMTI Institute for Biomedical Technology.

References

[1] Perfusion In Burn Injuries Rapidly Determined By Using Improved Laser-Doppler Technology, Hospital Test Shows. *Science Daily*.

http://www.sciencedaily.com/releases/2007/12/071216130313.htm

[2] University of Twente (2007, December 16). Perfusion In Burn Injuries Rapidly Determined By Using Improved Laser-Doppler Technology, Hospital Test Shows. *ScienceDaily*. Retrieved January 20, 2012, from http://www.sciencedaily.com /releases/2007/12/071216130313.htm

Doppler Radars Help Increase Monsoon Rainfall Prediction Accuracy

ScienceDaily (Oct. 5, 2010) — Doppler weather radar will significantly improve forecasting models used to track monsoon systems influencing the monsoon in and around India, according to a research collaboration including Purdue University, the National Center for Atmospheric Research and the Indian Institute of Technology Delhi.

Dev Niyogi, a Purdue associate professor of agronomy and earth and atmospheric sciences, said modeling of a monsoon depression track can have a margin of error of about 200 kilometers for landfall, which can be significant for storms that produce as much as 20-25 inches of rain as well as inland floods and fatalities.

"When you run a forecast model, how you represent the initial state of the atmosphere is critical. Even if Doppler radar information may seem highly localized, we find that it enhances the regional atmospheric conditions, which, in turn, can significantly improve the dynamic prediction of how the monsoon depression will move as the storm makes landfall," Niyogi said. "It certainly looks like a wise investment made in Doppler radars can help in monsoon forecasting, particularly the heavy rain from monsoon processes."

Niyogi, U.C. Mohanty, a professor in the Centre for Atmospheric Sciences at the Indian Institute of Technology, and Mohanty's doctoral student, Ashish Routray, collaborated with scientists at the National Center for Atmospheric Research and gathered information such as radial velocity and reflectivity from six Doppler weather radars that were in place during storms. Using the Weather Research and Forecasting Model, they found that incorporating the Doppler radar-based information decreased the error of the monsoon depression's landfall path from 200 kilometers to 75 kilometers.

Monsoons account for 80 percent of the rain India receives each year. Mohanty said more accurate predictions could better prepare people for heavy rains that account for a number of deaths in a monsoon season.

"Once a monsoon depression passes through, it can cause catastrophic floods in the coastal areas of India," Mohanty said. "Doppler radar is a very useful tool to help assess these things."

The researchers modeled monsoon depressions and published their findings in the Quarterly Journal of the Royal Meteorological Society. Future studies will incorporate more simulations and more advanced models to test the ability of Doppler radar to track monsoon processes. Niyogi said the techniques and tools being developed also could help predict landfall of tropical storm systems that affect the Caribbean and the United States.

The National Science Foundation CAREER program, U.S. Agency for International Development and the Ministry of Earth Sciences in India funded the study.

References

[1] Doppler Radars Help Increase Monsoon Rainfall Prediction Accuracy. *Science Daily*.

http://www.sciencedaily.com/releases/2010/10/101005171044.htm

[2] Purdue University (2010, October 5). Doppler radars help increase monsoon rainfall prediction accuracy. *ScienceDaily*. Retrieved January 20, 2012, from http://www.sciencedaily.com/releases/2010/10/101005171044.htm

Explain: the Doppler Effect

The same phenomenon behind changes in the pitch of a moving ambulance's siren is helping astronomers locate and study distant planets.

Many students learn about the Doppler effect in physics class, typically as part of a discussion of why the pitch of a siren is higher as an ambulance approaches and then lower as the ambulance passes by. The effect is useful in a variety of different scientific disciplines, including planetary science: Astronomers rely on the Doppler Effect to detect planets outside of our solar system, or exoplanets. To date, 442 of the 473 known exoplanets have been detected using the Doppler Effect, which also helps planetary scientists glean details about the newly found planets.

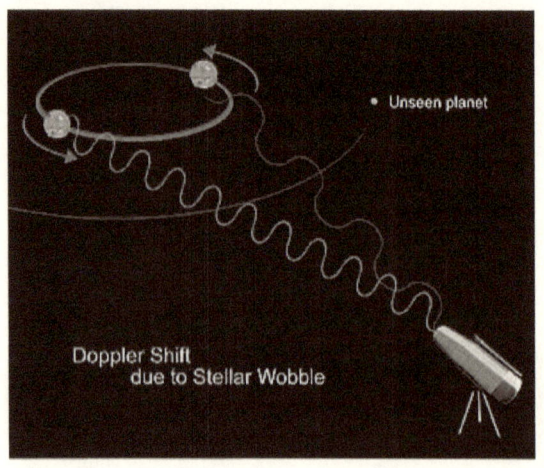

The Doppler Effect, or Doppler shift, describes the changes in frequency of any kind of sound or light wave produced by a moving source with respect to an observer. Waves emitted by an object traveling toward an observer get compressed — prompting a higher frequency — as the source approaches the observer. In contrast, waves emitted by a source traveling away from an observer get stretched out.

In astronomy, that source can be a star that emits electromagnetic waves; from our vantage point, Doppler shifts occur as the star orbits around its own center of mass and moves toward or away from Earth. These wavelength shifts can be seen in the form of subtle changes in its spectrum, the rainbow of colors emitted in light. When a star moves toward us, its wavelengths get compressed, and its spectrum becomes slightly bluer.

When the star moves away from us, its spectrum looks slightly reddened.

To observe the so-called red shifts and blue shifts over time, planetary scientists use a high-resolution prism-like instrument known as a spectrograph that separates incoming light waves into different colors. In every star's outer layer, there are atoms that absorb light at specific wavelengths, and this absorption appears as dark lines in the different colors of the star's spectrum that are recorded from the light emanating from the star. Researchers use the shifts in these lines as convenient markers by which to measure the size of the Doppler shift.

If the star exists by itself — that is, if there is no exoplanet or companion star in its stellar system — then there will be no change in the pattern of its Doppler shifts over time. But if there is a planet or companion star in the system, the gravitational pull of this unseen body or star will perturb the host star's movement at certain parts of its orbit, producing a noticeable change in the overall pattern and size of Doppler shifts over time. In other words, the pattern of a star's Doppler shifts can change over time as a result of gravity affecting the star's motion. "If this shift is large, then it must be caused by another star pulling it, but if this shift is small, then it is likely caused by a low-mass body like an exoplanet," explains Joshua Winn, an assistant professor in MIT's Department of Physics. As part of his work at MIT's Kavli Institute for Astrophysics and Space Research, Winn studies the relationship between an exoplanet's orbit and its parent star's rotation for clues about how the planet may have formed.

How a planet's Doppler shift changes over time can also shed light on the planet's orbital period (the length of its "year"), the shape of its orbit and its minimum possible mass. Recently, Kavli postdoc Simon Albrecht used the Doppler Effect to detect color shifts in the light absorbed by an exoplanet, which indicated strong winds in the planet's atmosphere.

Doppler shifts are used in many fields besides astronomy. By sending radar beams into the atmosphere and studying the changes in the wavelengths of the beams that come back, meteorologists use the Doppler Effect to detect water in the atmosphere. The Doppler phenomenon is also used in healthcare with echocardiograms that send ultrasound beams through a body to measure changes in blood flow to make sure that a heart valve is working properly or to diagnose vascular diseases. Police also rely on the Doppler Effect when they use a radar gun to bounce radio beams off of your car; the change in frequency between the directed and reflected beams provides a measure of your car's speed.

References

[1] Explain: the Doppler effect. MITnews (Massachusetts Institute of Technology).

http://web.mit.edu/newsoffice/2010/explained-doppler-0803.html

THE SONIC BOOM: A RELATED EFFECT.

A sonic boom is the sound associated with the shock waves created by an object traveling through the air faster than the speed of sound. Sonic booms generate enormous amounts of sound energy, sounding much like an explosion. The crack of a supersonic bullet passing overhead is an example of a sonic boom in miniature.

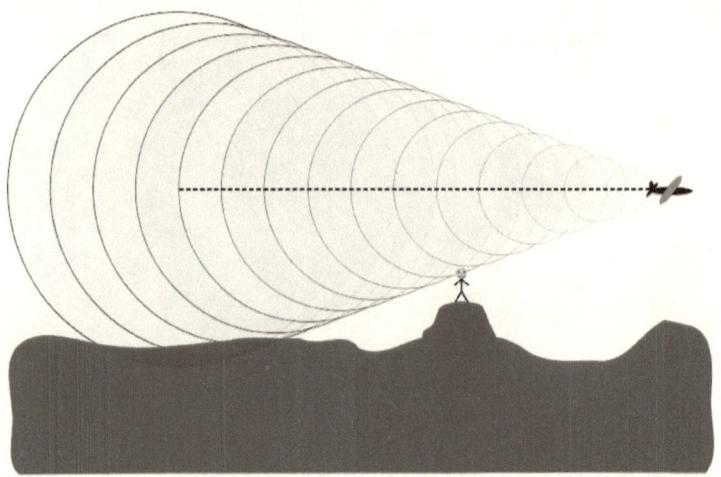

Some people today have had the experience of hearing a jet fly high overhead, producing a shock wave known as a sonic boom. A sonic boom, needless to say, is certainly not something of which Doppler would have had any knowledge, nor is it an illustration of the Doppler effect, per se. But it is an example of sound compression, and, therefore, it deserves attention here.

The speed of sound, unlike the speed of light, is dependent on the medium through which it travels. Hence, there is no such thing as a fixed "speed of sound"; rather, there is only a speed at which sound waves are transmitted through a given type of material. Its speed through a gas, such as air, is proportional to the square root of the pressure divided by the density. This, in turn, means that the higher the altitude, the slower the speed of sound: for the altitudes at which jets fly, it is about 660 MPH (1,622 km/h).

As a jet moves through the air, it too produces sound waves which compress toward the front, and widen toward the rear. Since sound waves themselves are really just fluctuations in pressure, this means that the faster a jet goes, the greater the pressure of the sound waves bunched up in front of it. Jet pilots speak of "breaking the sound barrier," which is more than just a figure of speech. As the craft approaches the speed of sound, the pilot becomes aware of a wall of high pressure to the front of the plane, and as a result of this high-pressure wall, the jet experiences enormous turbulence.

The speed of sound is referred to as Mach 1, and at a speed of between Mach 1.2 and Mach 1.4, even stranger things begin to happen. Now the jet is moving faster than the sound waves emanating from it, and, therefore, an observer on the ground sees the jet move by well before hearing the sound. Of course, this would happen to some extent anyway, since light travels so much faster than sound; but the difference between the arrival time of the light waves and the sound waves is even more noticeable in this situation.

Meanwhile, up in the air, every protruding surface of the aircraft experiences intense pressure: in particular, sound waves tend to become highly compressed along the aircraft's nose and tail.

Eventually these compressed sound waves build up, resulting in a shock wave. Down on the ground, the shock wave manifests as a "sonic boom"—or rather, two sonic booms—one from the nose of the craft, and one from the tail. People in the aircraft do not hear the boom, but the shock waves produced by the compressed sound can cause sudden changes in pressure, density, and temperature that can pose dangers to the operation of the airplane.

To overcome this problem, designers of supersonic aircraft have developed planes with wings that are swept back, so they fit within the cone of pressure.

When an object passes through the air it creates a series of pressure waves in front of it and behind it, similar to the bow and stern waves created by a boat. These waves travel at the speed of sound, and as the speed of the object increases, the waves are forced together, or compressed, because they cannot get out of the way of each other, eventually merging into a single shock wave at the speed of sound. This critical speed is known as Mach 1 and is approximately 1,225 km/h (761 mph) at sea level and 20 °C (68 °F). In smooth flight, the shock wave starts at the nose of the aircraft and ends at the tail. Because radial directions around the aircraft's direction of travel are equivalent, the shock forms a Mach cone with the aircraft at its tip. The half-angle (between direction of flight and the shock wave) α is given by:

$$\sin(\alpha) = \frac{v_{sound}}{v_{object}}$$

There is a rise in pressure at the nose, decreasing steadily to a negative pressure at the tail, followed by a sudden return to normal pressure after the object passes. This "overpressure profile" is known as an N-wave because of its shape. The "boom" is experienced when there is a sudden change in pressure, so the N-wave causes two booms, one when the initial pressure rise from the nose hits, and another when the tail passes and the pressure suddenly returns to normal. This leads to a distinctive "double boom" from supersonic aircraft. When maneuvering, the pressure distribution changes into different forms, with a characteristic U-wave shape.

Since the boom is being generated continually as long as the aircraft is supersonic, it fills out a narrow path on the ground following the aircraft's flight path, a bit like an unrolling red carpet and hence known as the "boom carpet". Its width depends on the altitude of the aircraft. The distance from the point on the ground where the boom is heard to the aircraft depends on its altitude and the angle α.

For today's supersonic aircraft in normal operating conditions, the peak overpressure varies from less than 50 to 500 Pa (one pound per square foot to about 10 pounds per square foot) for a N-wave boom. Peak overpressures for U-waves are amplified two to five times the N-wave, but this amplified overpressure impacts only a very small area when compared to the area exposed to the rest of the sonic boom.

The strongest sonic boom ever recorded was 7,000 Pa (144 pounds per square foot) and it did not cause injury to the researchers who were exposed to it. The boom was produced by a F-4 flying just above the speed of sound at an altitude of 100 feet (30 m). In recent tests, the maximum boom measured during more realistic flight conditions was 1,010 Pa (21 pounds per square foot).

There is a probability that some damage — shattered glass for example — will result from a sonic boom. Buildings in good repair should suffer no damage by pressures of 11 pounds per square foot or less. And, typically, community exposure to sonic boom is below two pounds per square foot. Ground motion resulting from sonic boom is rare and is well below structural damage thresholds accepted by the U.S. Bureau of Mines and other agencies.

The power, or volume, of the shock wave is dependent on the quantity of air that is being accelerated, and thus the size and shape of the aircraft. As the aircraft increases speed the shock cone gets tighter around the craft and becomes weaker to the point that at very high speeds and altitudes no boom is heard. The "length" of the boom from front to back is dependent on the length of the aircraft to a factor of 3:2.

Longer aircraft therefore "spread out" their booms more than smaller ones, which leads to a less powerful boom.

Several smaller shock waves can, and usually do, form at other points on the aircraft, primarily any convex points or curves, the leading wing edge and especially the inlet to engines.

These secondary shockwaves are caused by the air being forced to turn around these convex points, which generates a shock wave in supersonic flow.

The later shock waves are somewhat faster than the first one, travel faster and add to the main shockwave at some distance away from the aircraft to create a much more defined N-wave shape. This maximizes both the magnitude and the "rise time" of the shock which makes the boom seem louder. On most designs the characteristic distance is about 40,000 feet (12,000 m), meaning that below this altitude the sonic boom will be "softer". However, the drag at this altitude or below makes supersonic travel particularly inefficient, which poses a serious problem.

Stationary Sound Source

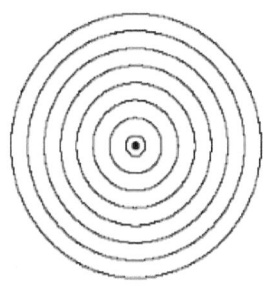

The photo at left shows a stationary sound source. Sound waves are produced at a constant frequency f_0, and the wavefronts propagate symmetrically away from the source at a constant speed v, which is the speed of sound in the medium. The distance between wavefronts is the wavelength. All observers will hear the same frequency, which will be equal to the actual frequency of the source.

Source moving with $v_{source} < v_{sound}$ (Mach 0.7)

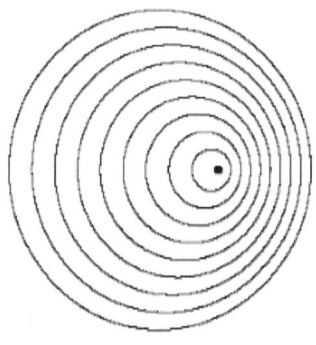

In the photo at left the same sound source is radiating sound waves at a constant frequency in the same medium. However, now the sound source is moving to the right with a speed $v_s = 0.7\ v$ (Mach 0.7).

The wavefronts are produced with the same frequency as before. However, since the source is moving, the center of each new wavefront is now slightly displaced to the right.

As a result, the wavefronts begin to bunch up on the right side (in front of) and spread further apart on the left side (behind) of the source. An observer in front of the source will hear a higher frequency $f' > f_0$, and an observer behind the source will hear a lower frequency $f' < f_0$.

Source moving with $v_{source} = v_{sound}$ (Mach 1 - breaking the sound barrier)

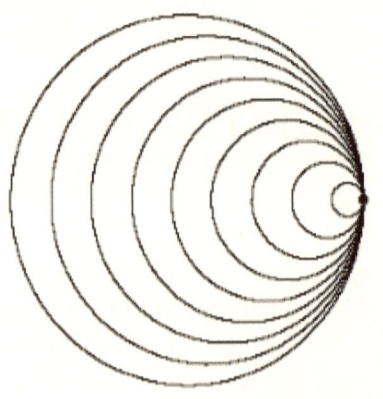

Now the source is moving at the speed of sound in the medium ($v_s = v$, or Mach 1). The speed of sound in air at sea level is about 340 m/s or about 750 mph. The wavefronts in front of the source are now all bunched up at the same point. As a result, an observer in front of the source will detect nothing until the source arrives. The

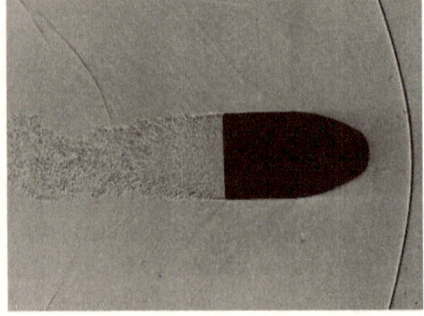

pressure front will be quite intense (a shock wave), due to all the wavefronts adding together, and will not be percieved as a pitch but as a "thump" of sound as the pressure wall passes by. The figure at right shows a bullet travelling at Mach 1.01. You can see the shock wave front just ahead of the bullet.

. Jet pilots flying at Mach 1 report that there is a noticeable "wall" or "barrier" which must be penetrated before achieving supersonic speeds. This "wall" is due to the intense pressure front, and flying within this pressure front produces a very turbulent and bouncy ride. Chuck Yeager was the first person to break the sound barrier when he flew faster than the speed of sound in the X-1 rocket-powered aircraft on October 14, 1947. Check out the movie The Right Stuff for more about this significant milestone, and the beginnings of the US space project.

The figure at right shows a F-18 at the exact instant it goes supersonic. Click on the figure to see more information and a MPEG movie of this event.

Source moving with $v_{source} > v_{sound}$ (Mach 1.4 - supersonic)

The sound source has now broken through the sound speed barrier, and is traveling at 1.4 times the speed of sound (Mach 1.4). Since the source is moving faster than the sound waves it creates, it actually leads the advancing wavefront. The sound source will pass by a stationary observer before the observer actually hears the sound it creates.

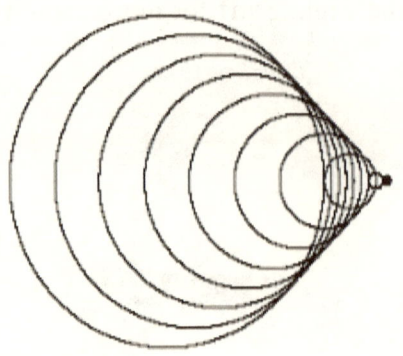

Sonic Booms

A sonic boom is the thunder-like noise a person on the ground hears when an aircraft or other type of aerospace vehicle flies overhead faster than the speed of sound or supersonic.

Air reacts like a fluid to supersonic objects. As objects travel through the air, the air molecules are pushed aside with great force and this forms a shock wave much like a boat creates a bow wave. The bigger and heavier the aircraft, the more air it displaces.

The Cause

The shock wave forms a cone of pressurized air molecules which move outward and rearward in all directions and extend to the ground. As the cone spreads across the landscape along the flight path, they create a continuous sonic boom along the full width of the cone's base. The sharp release of pressure, after the buildup by the shock wave, is heard as the sonic boom.

The change in air pressure associated with a sonic boom is only a few pounds per square foot -- about the same pressure change experienced riding an elevator down two or three floors. It is the rate of change, the sudden onset of the pressure change that makes the sonic boom audible.

All aircraft generate two cones, at the nose and at the tail. They are usually of similar strength and the time interval between the two as they reach the ground is primarily dependent on the size of the aircraft and its altitude. Most people on the ground cannot distinguish between the two and they are usually heard as a single sonic boom. Sonic booms created by vehicles the size and mass of the space shuttle are very distinguishable and two distinct booms are easily heard.

General Factors Associated With Sonic Booms

There are several factors that can influence sonic booms -- weight, size, and shape of the aircraft or vehicle, plus its altitude, attitude and flight path, and weather or atmospheric conditions.

A larger and heavier aircraft must displace more air and create more lift to sustain flight, compared with small, light aircraft. Therefore, they will create sonic booms stronger and louder than those of smaller, lighter aircraft. The larger and heavier the aircraft, the stronger the shock waves will be.

Altitude determines the distance shock waves travel before reaching the ground, and this has the most significant effect on intensity. As the shock cone gets wider, and it moves outward and downward, its strength is reduced. Generally, the higher the aircraft, the greater the distance the shock wave must travel, reducing the intensity of the sonic boom. Of all the factors influencing sonic booms, increasing altitude is the most effective method of reducing sonic boom intensity.

The width of the boom "carpet" beneath the aircraft is about one mile for each 1000 feet of altitude. An aircraft, for example, flying supersonic at 50,000 feet can produce a sonic boom cone about 50 miles wide. The sonic boom, however, will not be uniform. Maximum intensity is directly beneath the aircraft, and decreases as the lateral distance from the flight path increases until it ceases to exist because the shock waves refract away from the ground. The lateral spreading of the sonic boom depends only upon altitude, speed and the atmosphere -- and is independent of the vehicle's shape, size, and weight.

The ratio of aircraft length to maximum cross sectional area also influences the intensity of the sonic boom. The longer and more slender the aircraft, the weaker the shock waves.

The fatter and more blunt the vehicle, the stronger the shock wave can be.

Increasing speeds above Mach 1.3 results in only small changes in shock wave strength.

The direction of travel and strength of shock waves are influenced by wind, speed, and direction, and by air temperature and pressure. At speeds slightly greater than Mach 1, their effect can be significant, but their influence is small at speeds greater than Mach 1.3. Distortions in the shape of the sonic boom signatures can also be influenced by local air turbulence near the ground. This, too, will cause variations in the overpressure levels.

Aircraft maneuvering can cause distortions in shock wave patterns. Some maneuvers -- pushovers, acceleration and "S" turns -- can amplify the intensity of the shock wave. Hills, valleys and other terrain features can create multiple reflections of the shock waves and affect intensity.

Overpressure

Sonic booms are measured in pounds per square foot of overpressure. This is the amount of the increase over the normal atmospheric pressure which surrounds us (2,116 psf/14.7 psi).

At one pound overpressure, no damage to structures would be expected.

Overpressures of 1 to 2 pounds are produced by supersonic aircraft flying at normal operating altitudes. Some public reaction could be expected between 1.5 and 2 pounds.

Rare minor damage may occur with 2 to 5 pounds overpressure.

As overpressure increases, the likelihood of structural damage and stronger public reaction also increases. Tests, however, have shown that structures in good condition have been undamaged by overpressures of up to 11 pounds.

Sonic booms produced by aircraft flying supersonic at altitudes of less than 100 feet, creating between 20 and 144 pounds overpressure, have been experienced by humans without injury.

Damage to eardrums can be expected when overpressures reach 720 pounds. Overpressures of 2160 pounds would have to be generated to produce lung damage.

Typical overpressure of aircraft types are:

- **SR-71:** 0.9 pounds, speed of Mach 3, 80,000 feet

- **Concorde SST:** 1.94 pounds, speed of Mach 2, 52,000 feet

- **F-104:** 0.8 pounds, speed of Mach 1.93, 48,000 feet

- **Space Shuttle:** 1.25 pounds, speed of Mach 1.5, 60,000 feet, landing approach

SEE YOU SOON!

www.ingramcontent.com/pod-product-compliance
Lightning Source LLC
Chambersburg PA
CBHW021019180526
45163CB00005B/2019